BUILD YOUR OWN
AMAZING
SPACE EXPERIMENTS

Rob Ives

CONTENTS

PREPARE FOR LIFTOFF	4	SAFETY FIRST!	6

AMAZING SPACE EXPERIMENTS	8	SUPER SOLAR SYSTEM MODEL	21
ORBITTING ORERRY	10	ROCKET POWER	24
SOALR ECLIPSE	14	SPACE EXPERIMENTS THROUGH TIME	26
EGG-CELLENT PARACHUTE	16	AMAZING ASTRONAUTS	28
WEIGHT SLIDE RULER	18	EXTRA-TERRESTRIAL EXPERIMENTS	30
VACUUM EXPERIMENT	20		
		GLOSSARY AND INDEX	32

Copyright © 2024 Hungry Tomato Ltd

First published in 2024 by Hungry Tomato Ltd
F15, Old Bakery Studios, Blewetts Wharf, Malpas Road, Truro, Cornwall,
TR1 1QH, UK.

No part of this publication may be reproduced, stored in a retrieval system, or transmitted in any form or by any means, electronic, mechanical, photocopying, recording, or otherwise, without prior written permission of the copyright owner.
A CIP catalogue record for this book is available from the British Library.

ISBN 9781916598867

Printed in China

Discover more at
www.hungrytomato.com

PREPARE FOR LIFTOFF

Try your hand at building amazing space-themed models! Using smart and simple engineering principles, you can make a whole collection of out-of-this-world crafts that hover, fly, move, and show the wonders of our universe and beyond!

THIS BOOK IS INTERACTIVE!

Some of the projects in this book come with templates to help you cut pieces to the right shape and size. Use a smartphone to scan the QR code at the beginning of the project to access a downloadable template that you can print out.

You will find QR codes at the end of some projects, too. These will direct you to videos of the moving models in action!

You can also find all templates and videos at:
www.hungrytomato.com/space-experiments

PREPARE FOR LIFTOFF

TOP TIPS

- Before you start any project, read the step-by-steps all the way through to get an idea of what you are aiming for. The pictures show what the steps tell you.

- When printing templates, check that your printer is set to "print to scale" or to "full size" to make sure they come out the right size for your other materials!

- Use a cutting mat, or similar surface, for cutting lengths of craft sticks, skewers, and anything else you may need.

- Use the sharp end of a pencil to make small holes in cardboard (see page 7 for method) or ask an adult to help, using either scissors or a craft knife.

- Ask an adult to help straighten out and shape paper clips using a pair of pliers.

- Where strong glue is required, you may want to use a glue gun. Make sure you ask permission, and do not use it without an adult present. Strong liquid glue, such as wood or epoxy glue, will work well, too.

⭐ EASY

⭐⭐ MEDIUM

⭐⭐⭐ HARDER

You will find stars in the corner of the first page of each craft. These stars are a guide to the difficulty level of each project. They show you when you may need another pair of hands!

SAFETY FIRST!

Be careful and use good sense when making these models. They are easy to understand but will require some cutting, gluing, drilling, and other awkward tasks that you may need some help with from an adult.

WHEN TO GET HELP

Watch out for this sign throughout the book. You may need help from an adult when completing these tasks.

DISCLAIMER

The author, publisher, and bookseller cannot take responsibility for your safety. When you make and try out the projects, you do so at your own risk. Look out for the safety warning symbol (shown left) given throughout the book and call on adult assistance when you are cutting materials or using a pair of scissors or pliers, craft drill, or hot glue.

HOW TO CUT A POSTER TUBE SAFELY:

1.
Cut a strip of cardboard and fold a right angle into it. Measure from the crease the width you need the tube to be and make a hole at that point.

2.
Hold the cardboard over the end of the poster tube. With a pencil in the hole, twist the tube around to draw a line parallel to the edge.

3.
Ask an adult to carefully cut along the line to make a short section of tube. They could use scissors or a craft knife.

SAFETY FIRST!

HOW TO SCORE PAPER OR CARD SAFELY:

Using a ruler as a guide, press along the line with a hard plastic item, like the end of a pen lid. Do this on a cutting mat to protect the surface you're working on.

HOW TO MAKE HOLES IN CARDBOARD SAFELY AND EASILY:

Pressing a pencil point through cardboard and into an eraser, like the photo on the right, is a safe and easy way to make holes.

AMAZING SPACE EXPERIMENTS

The first astronomers came up with ideas about outer space. They didn't have the right equipment to put them to the test, but now modern scientists do!

LEARNING ABOUT PLANETS
For years, **astronomers** could only do experiments on Earth. They created models, like the Orbiting Orrery (page 10), to demonstrate the way that the planets move around the Sun. These scientific models help us understand what's happening in our solar system.

THE BIG QUESTIONS
Gravity was discovered in the 1600s when scientist Sir Isaac Newton questioned why an apple falls to the ground, rather than floating upward or sideways! His work was revolutionary as it helped explain why the Moon keeps **orbiting** the Earth, and Earth keeps orbiting the Sun.

WANT TO KNOW MORE?
This book is full of amazing experiments. From watching how astronauts land back on Earth with an exciting eggs-periment to testing the power of a vacuum, there's lots of fun projects to try!

ORBITING ORRERY

Make this awesome moving model to see how Earth and its Moon orbit the Sun!

Use the QR code to access the template you need.

WHAT YOU NEED:
- Paper straw 6mm
- Paper straw 8mm
- Corrugated cardboard
- Paper clip
- Single-sided corrugated paper
- Pearl-like bead 6mm
- Blue and yellow paint
- Wooden bead or sculpting clay 15mm
- Table tennis ball
- 2 wooden skewers

TOOLS:
- Pair of scissors
- Strong craft glue
- Black marker pen
- Needle pliers
- Paper packing tape
- Pencil and eraser

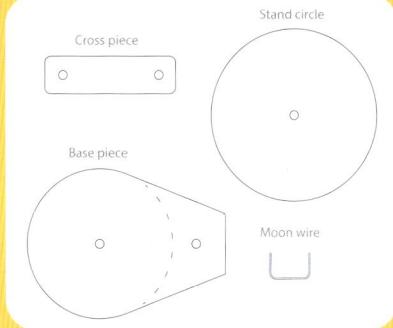

1 Print, copy, or trace the shapes from the template onto corrugated cardboard and cut out. Make holes where indicated.

Don't worry if Earth slides down the skewer. This will be fixed in step 6.

This will represent the Earth

2 Trim two wooden skewers to 100mm long. Fit the wooden bead or sculpting clay onto one of the skewers and paint blue.

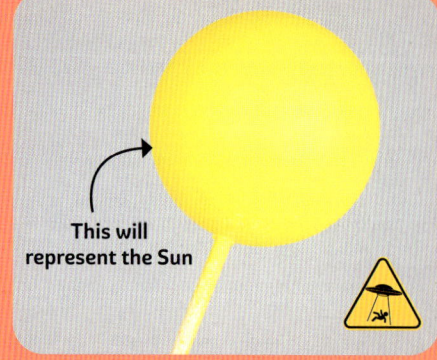

This will represent the Sun

3 Ask an adult to make a small hole in the table tennis ball. Then, slide the second skewer into the hole. Paint the ball yellow.

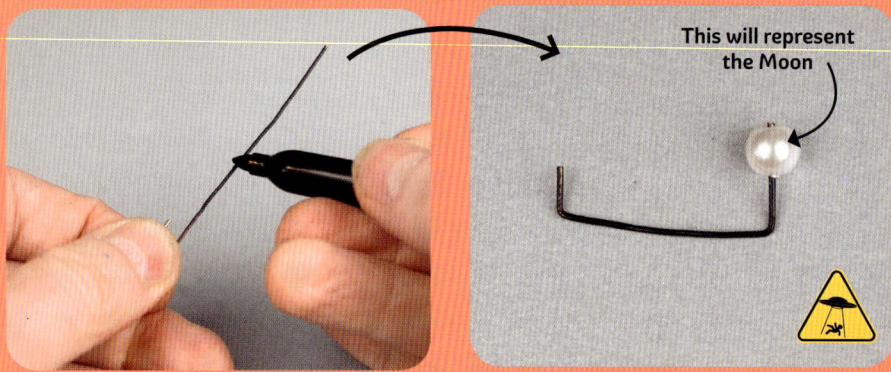

This will represent the Moon

4 Uncoil a paper clip, then shade it in with black marker pen. Ask an adult to bend the paper clip into the shape shown using pliers. Fit the pearl-like bead on top, gluing into place.

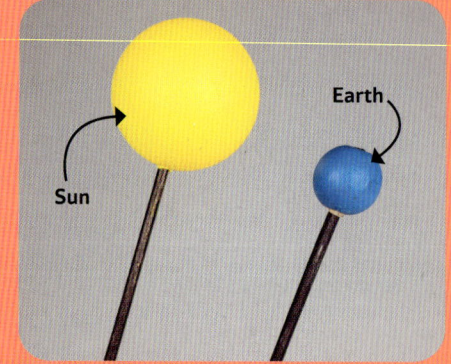

Sun — Earth

5 Shade in the Sun and Earth skewers with black marker pen.

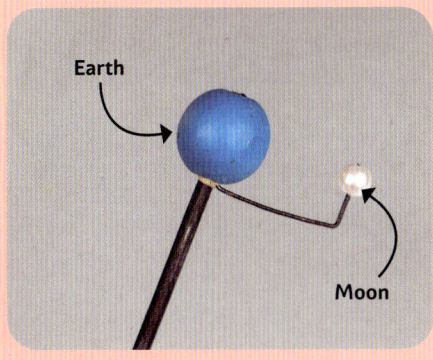

6 Make sure the Earth is at the top of the skewer, then slot the Moon's paper clip arm inside the bead. Glue both in place.

7 Make a small hole using a pencil and eraser (see page 7) in the middle of each corrugated cardboard circle from the template pieces. The hole should be just big enough for one of the narrower straws to fit inside.

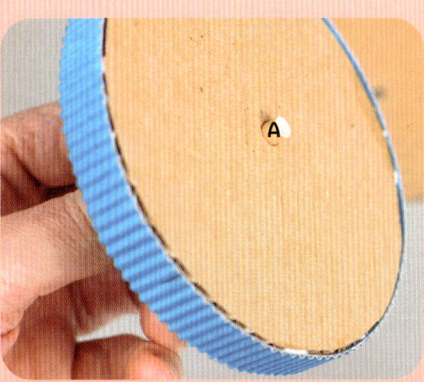

8 Thread the circles onto straw A (6mm) so that they line up. Then, glue the circles together.

9 Cut a piece of single-sided corrugated paper to fit perfectly around the circular cardboard stand. Glue in place as shown.

10 Push straw A further into the stand so that it pokes out the same amount as the height of the base piece that you cut out from the template.

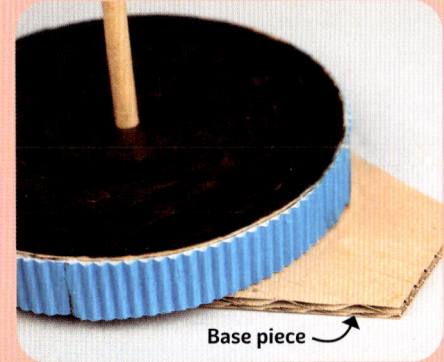
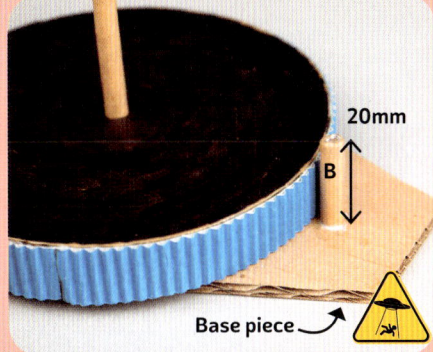
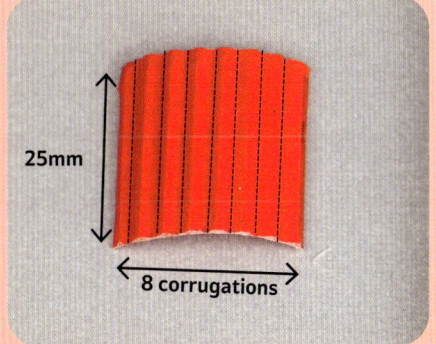

11 Slot the stand into the base where indicated on the template. It should still be able to turn freely. Decorate the stand with black marker.

12 Cut the other 6mm straw (straw B) to 20mm high. Then, glue it to the base where indicated on the template.

13 Cut a strip of single-sided corrugated paper to 25mm high and 8 corrugations wide.

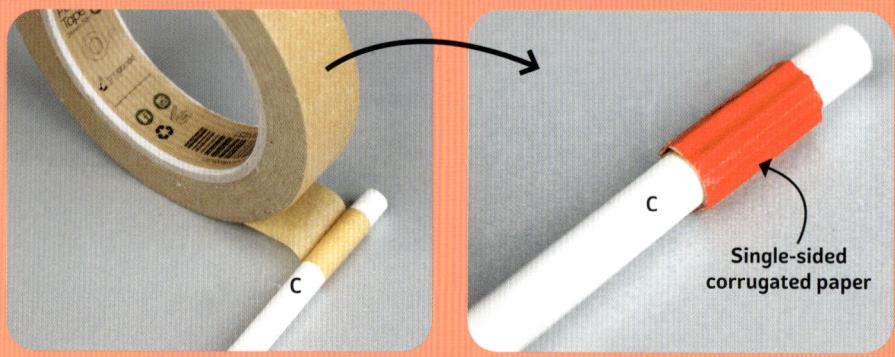 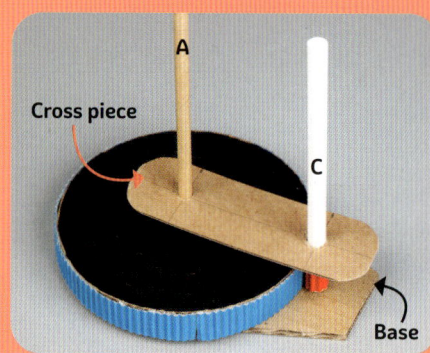

14 Roll tape around straw C (8mm) until it's thick enough so that the single-sided corrugated paper wraps exactly around and the ends meet up. Glue together.

15 Slot straw C onto straw B, with the corrugated card at the bottom. Slide the cross piece from the template onto the straws as shown.

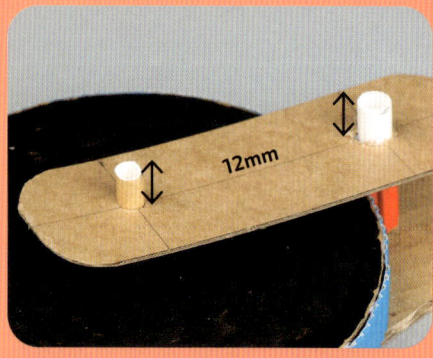 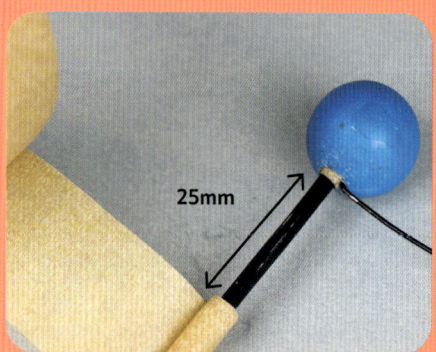 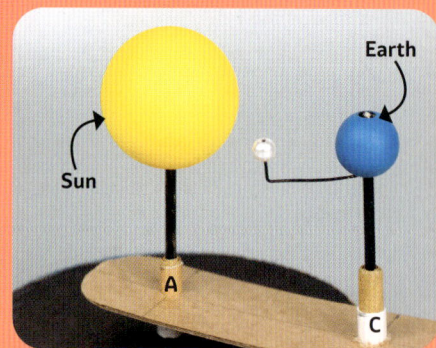

16 Trim straws A and C so that 12mm remains above the cross piece.

17 Tape around the Earth and Sun skewers so that when they're placed in the straws they stand upright.

18 Place the Earth and Moon skewer into straw C, and the Sun into straw A.

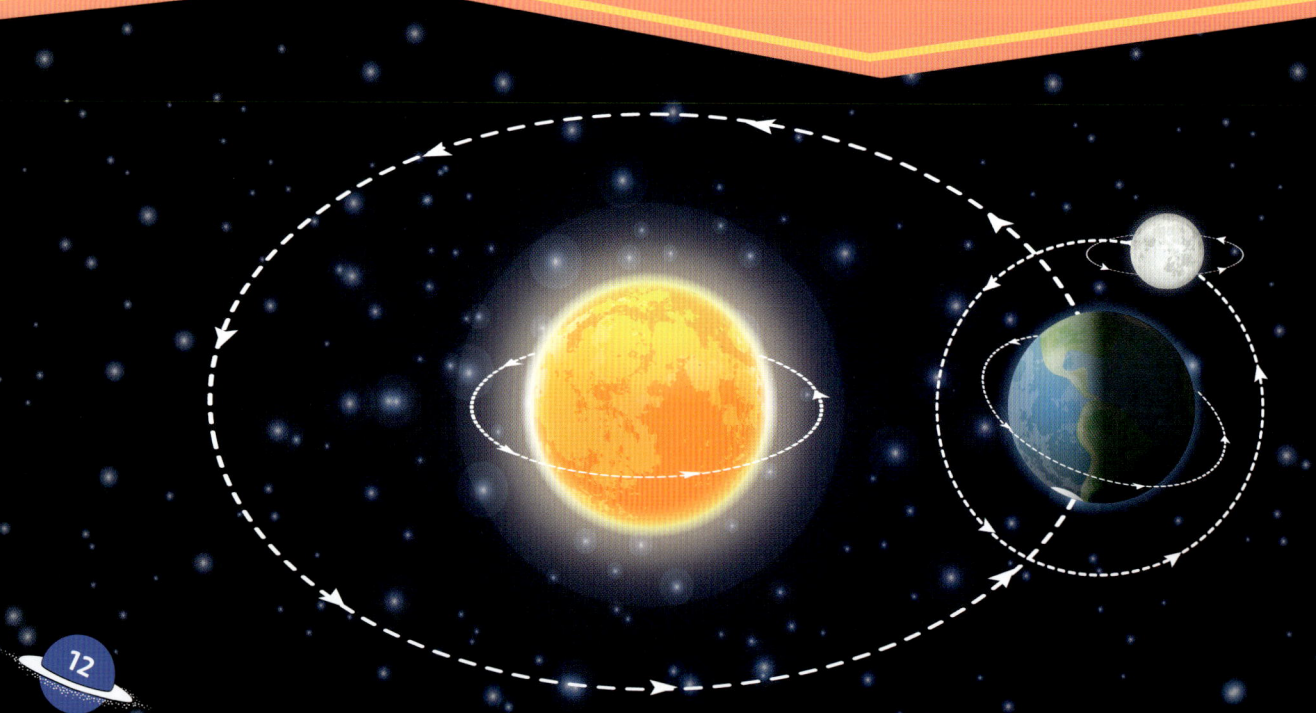

Guide piece

HOLD THE GUIDE PIECE AND TWIST TO SEE THE MOON ORBIT EARTH!

DID YOU KNOW?
In reality, it takes Earth 365 days - a whole year - to complete a full orbit of the Sun. During that time, the planet travels over 584 million miles! Our Moon follows, constantly spinning around Earth, too. What a journey!

SOLAR ECLIPSE

A solar eclipse happens when the Moon moves between Earth and the Sun, gradually blocking out the light from the Sun. You can understand how that is possible with this cool model!

WHAT YOU NEED:
- Table tennis ball
- Yellow and black paint
- Wooden bead or sculpting clay 12mm across
- Bamboo skewers
- Corrugated cardboard
- Felt-tip pens or pencils (optional)

TOOLS:
- Ruler
- Pair of scissors
- Strong craft glue

This will represent the Sun

1 Ask an adult to make a small hole in the table tennis ball. Slide a wooden skewer into the hole. Paint the ball yellow.

This will represent the Moon

2 Fit the bead or sculpting clay onto a wooden skewer. Don't worry if it slides down - we will fix this in step 3. Paint the bead black.

3 Once both the Sun and Moon are dry, replace the skewers with clean ones. Glue in place to secure.

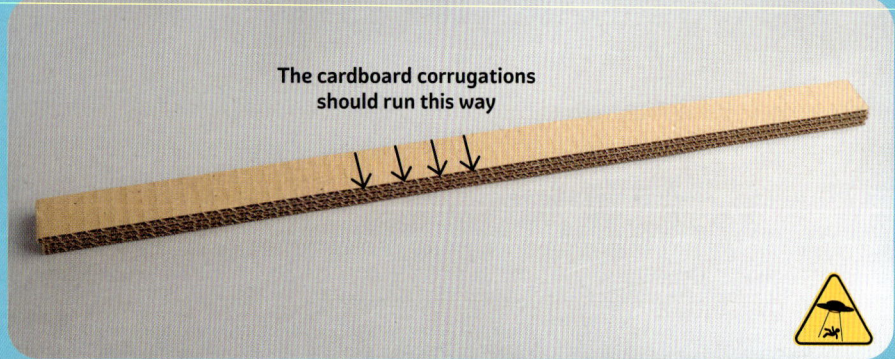

The cardboard corrugations should run this way

4 Carefully cut 3 strips of corrugated cardboard to 25mm x 450mm with the corrugations running in the direction shown. Glue the cardboard strips together, making one thick strip.

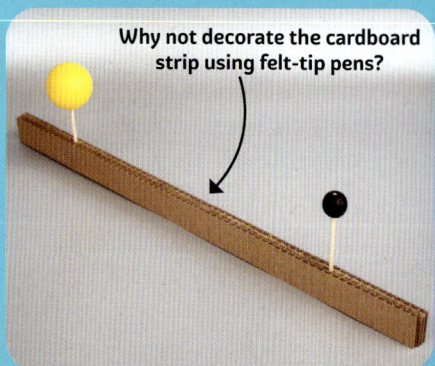

Why not decorate the cardboard strip using felt-tip pens?

5 Turn the cardboard strip to stand on its long, thin edge. Fit the Sun at one end and the Moon at the other.

How to use your solar eclipse model:

1.
Position your eye (or camera) in line with the model.

2.
With the Moon closest to you, move your head (or camera) right to left to see the phases of an eclipse.

This is a partial eclipse
This is when the Moon covers part of the Sun.

This is a total eclipse
This is when the Moon completely covers the Sun.

DID YOU KNOW?
The Moon doesn't make its own light but reflects the light of the Sun. Why not paint your Moon half yellow and half black (vertically) to show which side of the Moon is facing the Sun?

EGG-CELLENT PARACHUTE

TOP TIP

Use a paper cup if you don't have a poster tube with a plastic end.

This experiment shows **aerodynamics** in action! One of the safest and easiest ways for astronauts to land on Earth is to use a parachute. Can you land this egg safely, too?

WHAT YOU NEED:
- Hard-boiled egg
- Thread/thin string
- Poster tube or paper cup
- Plastic bag
- Tissue or cotton wool balls

TOOLS:
- Pair of scissors
- Sticky tape or strong craft glue
- Ruler
- Pencil and eraser

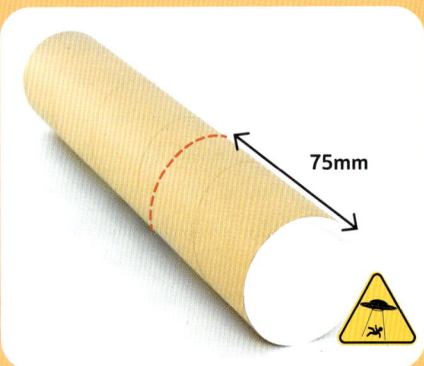

1 Cut the poster tube to a height of 75mm. Secure the plastic end with tape or glue to make sure the egg won't fall out!

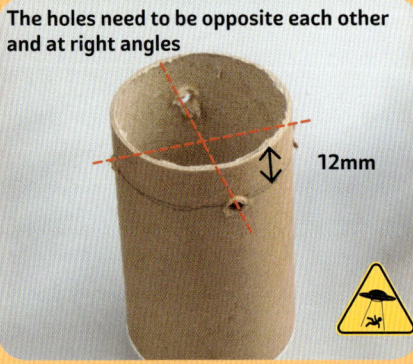

The holes need to be opposite each other and at right angles

2 Ask an adult to make four small holes (see page 7) about 12mm from the top of the tube.

3 Cut 4 lengths of thread to 460mm long.

4 Take one piece of thread and tie it to a hole in the poster tube. Repeat until each hole has a piece of thread tied to it. Place some tissue or cotton wool balls inside followed by the egg.

TOP TIP

Make sure you tie strong double knots in this craft to hold everything together. Ask an adult to help you to make sure they are secure!

5 Cut a piece of plastic bag roughly 460mm square to make the parachute.

The threads must not cross over each other

6 Pair up each piece of thread with one corner of the plastic bag. Tie the thread tightly to its corner.

HOW DOES THIS WORK?
As the egg is pulled toward the surface of the Earth by gravity, a strong force produced by **air resistance** pushes upward against the parachute. This reduces the force of gravity and slows the egg's fall.

7 Fill the top of the poster tube with tissue or cotton wool balls to keep the egg in place.

THROW THE PARACHUTE AS HIGH AS YOU CAN AND WATCH IT DRIFT SAFELY BACK DOWN TO EARTH!

WEIGHT SLIDE RULER

Did you know that light things, like this book, would weigh more on other planets? Use this fun slide ruler to work out weight across the solar system!

Use the QR code to access the template you need.

WHAT YOU NEED:
- Corrugated cardboard
- Thin card
- Plain white paper
- Clear plastic sheet
- Black marker pen

TOOLS:
- Ruler
- Pair of scissors
- Strong craft glue
- Sticky tape

1 Print, copy, or trace the shapes from the template onto the specified materials and cut out.

2 Glue the three paper parts from the template on top of their corresponding corrugated cardboard part.

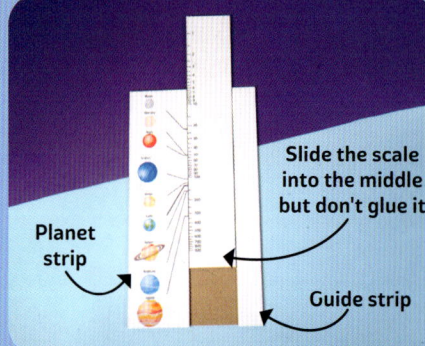

3 Glue the planet strips onto the backboard with the left sides lined up. Then, glue the guide strip onto the backboard with the right sides lined up.

4 To make the cursor, take the card cursor piece from the template. Score (see page 7) then fold along the dotted lines.

5 Cut out a piece of clear plastic using the cursor front template. Measure and mark a horizontal line through the middle with a black marker pen.

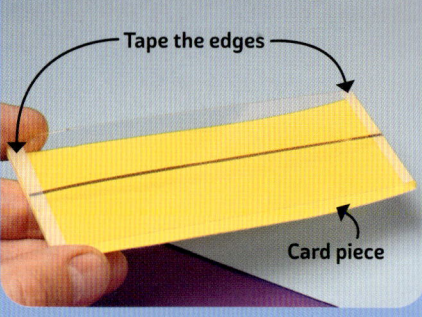

6 Balance the plastic on the folded sides of the card piece and tape the edges where they meet to complete the cursor.

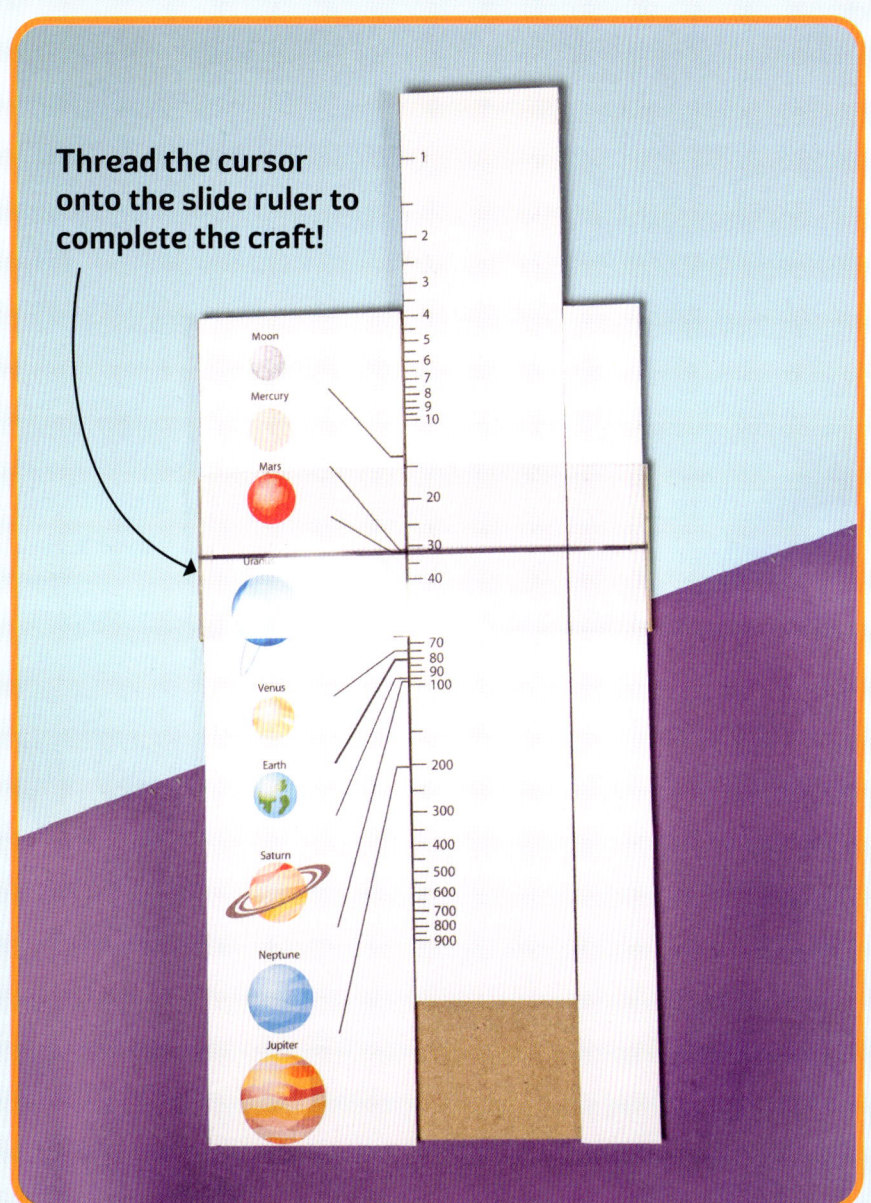

Thread the cursor onto the slide ruler to complete the craft!

How to use your weight slide ruler:

1.
Line up the scale so that Earth is pointing to the weight of the item.

2.
Without moving the scale, read what the weight would be on the other planets.

Example:
If something weighs 36kg on Earth, it would only weigh about 6kg on the Moon, but on Jupiter it would weigh over 90kg!

Have fun working out what different things would weigh on the different planets. Why not try the following:

- This book
- Shoe
- Phone
- Toy

DID YOU KNOW?
Weight is a combination of an item's mass (everything something's made of) and the strength of gravity on the planet (the force keeping your feet on the ground). An item's shape and size stay the same from planet to planet, but because each planet has a different gravity strength, it would appear lighter or heavier depending on where it is!

VACUUM EXPERIMENT

Outer space is a vacuum – this means there's no air! Create the vacuum of outer space down here on Earth with this extraordinary experiment.

WHAT YOU NEED:
- Large syringe with plastic tube
- Small glass jar
- Marshmallows

TOOLS:
- Pair of scissors
- Glue gun (cool melt) or strong craft glue

1 Ask an adult to make a hole in the lid of the jar - **don't do this yourself!** It needs to be big enough for the plastic tube to fit through.

2 Thread the end of the plastic tube through the hole and seal with glue.

The jar should be airtight

3 Fill the jar with marshmallows, then screw the lid on tightly. Attach the syringe to the tube.

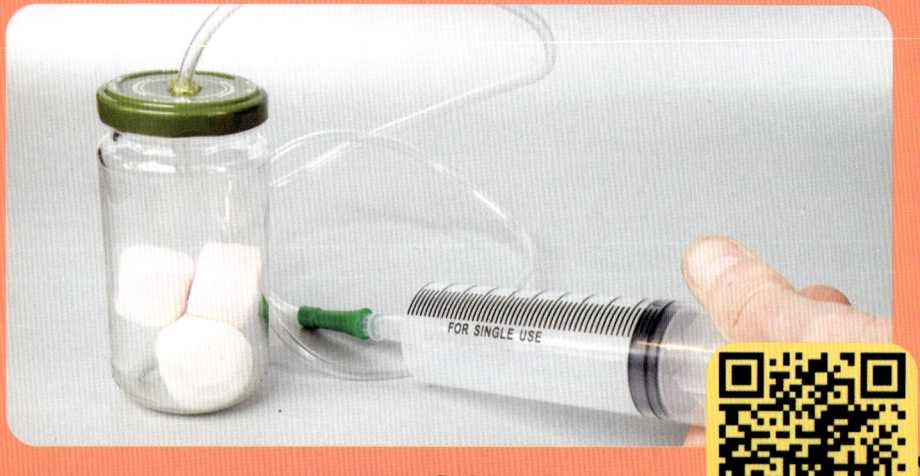

DID YOU KNOW?
By removing all the air from the jar, we create a vacuum. You can see what happens to the marshmallows with even a little of the air removed. Imagine what would happen to them in the vacuum of outer space!

PULL THE SYRINGE TO SEE THE MARSHMALLOWS EXPAND!

SUPER SOLAR SYSTEM MODEL

TOP TIP

Make up a funny rhyme to help you remember the order of the planets in our solar system.

Have fun making the solar system out of clay and see just how small Earth is compared to Jupiter!

WHAT YOU NEED:
- Ball 230mm across
- Assorted sculpting clay
- Small paper clips
- Assorted card

TOOLS:
- Plastic knife or sturdy ruler (for cutting clay)
- Strong craft glue
- Ruler
- Pair of scissors
- Needle pliers

JUPITER is identifiable by its banded cloud.

1 Roll out brown, yellow, and orange sculpting clay and stack on top of each other.

2 Slice off the edges of the pile to create a rough cylinder.

3 Shape the clay into a ball 24mm across.

SATURN is identifiable by its rings.

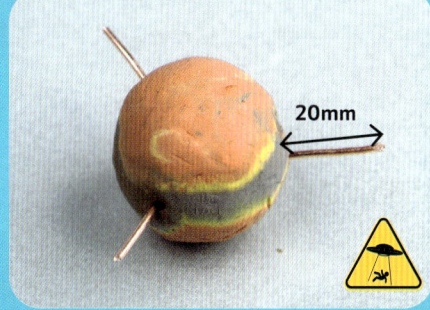
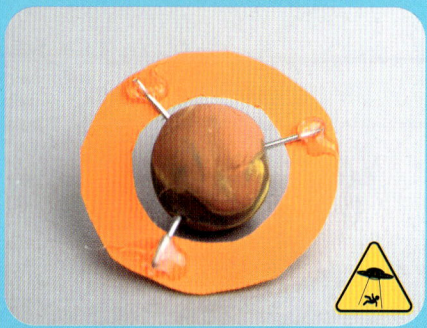

1 Make a brownish ball of sculpting clay 20mm across.

2 Straighten out three small paper clips and insert into the clay ball an equal distance apart.

3 Cut out a ring of card. Glue the ring to the paper clips as shown. Then, flip over to hide the ends.

HAVE FUN MAKING THE OTHER PLANETS!

Use this helpful chart to make sure your planets are to scale with each other:

PLANET	CLAY	SIZE
Mercury	Brown and yellow	1mm
Venus	Yellow and orange	2mm
Earth	Blue, green, and white	2mm
Mars	Red	1mm
Jupiter	Brown, yellow, and orange	24mm
Saturn	Brown and orange	20mm
Uranus	Blue (light)	7mm
Neptune	Blue (dark)	7mm

Can you name these planets in order, based on what you've learned?

Use a basketball (230mm) as the Sun when scaling up your planets.

 **TOP TIP** Why not make a stand for your solar system to display it? Fold a piece of cardboard on a right angle, paint it black, and decorate with stars. Then, place your planets on top. **Scan the QR code to download and print the planet names to glue on, too!**

DID YOU KNOW?
All the planets in our solar system take a different amount of time to orbit the Sun. The closest to the Sun, Mercury, takes 88 Earth days. The furthest, Neptune, takes 60,190 Earth days - that's 165 Earth years!

ROCKET POWER

Watch this car shoot off as the balloon loses air. This exciting experiment uses the same principle as rocket power: a **thrust** backward creates a surge forward.

WHAT YOU NEED:
- A balloon
- Marker pen 12mm wide
- 4 wooden beads 12mm
- 2 wooden skewers
- Poster tube 50mm in diameter
- 4 circular plastic lids 75mm in diameter
- 2 cable ties

TOOLS:
- Ruler
- Pencil
- Eraser
- Pair of scissors
- Sticky tape
- Craft knife

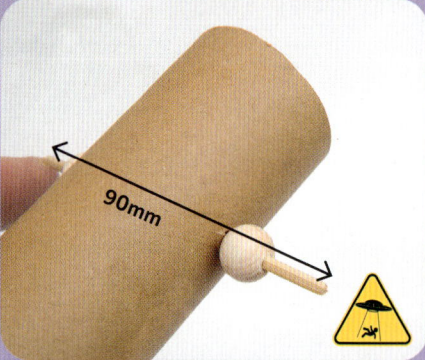

1 Ask an adult to cut a poster tube to 200mm long (see page 6) and make holes slightly larger than the skewers on opposite sides, about 40mm from each end. If these holes are too small, your car won't move very well. Don't worry; you can test this out in step 4.

2 Cut the skewers to 90mm, then thread them through the holes in the tube. Add a bead on each side as a washer.

TOP TIP

The plastic lid from crisp containers are the perfect size for the wheels! Better start eating; you need four lids!

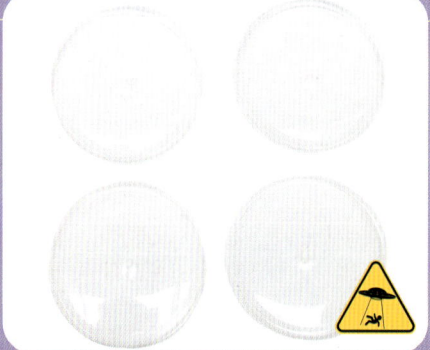

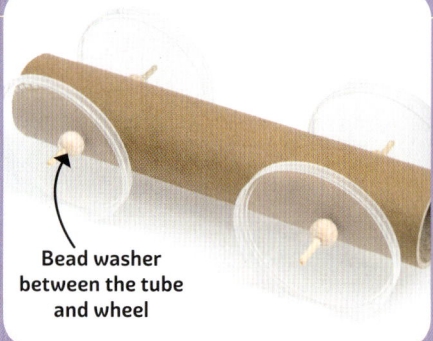

Bead washer between the tube and wheel

3 Ask an adult to make a small hole in the middle of each lid to fit the skewer through tightly.

4 Fit the lid wheels onto the skewers, with the rims facing outward. Gently push the car to make sure the wheels move smoothly.

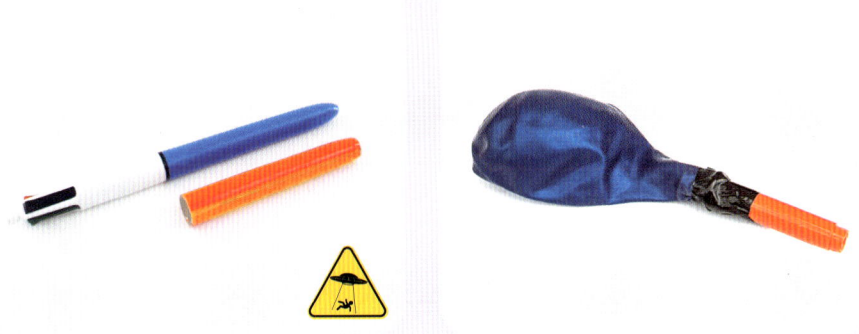
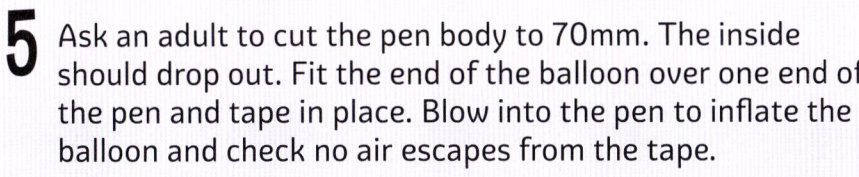
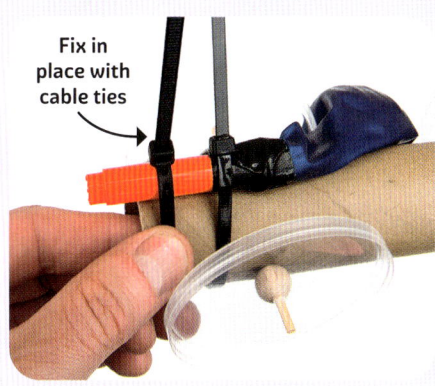

Fix in place with cable ties

5 Ask an adult to cut the pen body to 70mm. The inside should drop out. Fit the end of the balloon over one end of the pen and tape in place. Blow into the pen to inflate the balloon and check no air escapes from the tape.

6 Position the pen on the poster tube so that it extends beyond the end by 12mm.

BLOW THE BALLOON UP, PINCH IT ABOVE THE PEN, PLACE IT DOWN, AND RELEASE!

SPACE EXPERIMENTS THROUGH TIME

Our understanding of space has grown so much since we began journeying into outer space. You could say that space flight is the biggest space experiment of all time! How did it happen?

TESTING SPACE FLIGHT
It took lots of testing to make a vehicle that could travel fast and high enough to make space travel possible. The biggest discovery happened during the early 1900s when an American engineer, Robert Goddard, built the world's first liquid-fuelled rocket. Once scientists had tested this new technology on unmanned rockets, the race was on to send humans into space!

THE SPACE RACE
Our success at journeying to space is mostly thanks to the Space Race which was a big competition between the 1950s and 1970s when several countries wanted to be the first to succeed at space flight. Lots of big milestones were reached in a short space of time as scientists and engineers invented bigger and better spacecraft.

INTO THE FUTURE...
In 2012, scientists broke another record when Voyager 1, a **space probe** flew into **interstellar space** for the first time. This place at the edge of our solar system is the furthest that any human-made object has ever flown! Voyager 1 was first launched in 1977. It's had a long journey to get where it is today!

HUGE MOMENTS IN SPACE HISTORY

1957
A dog called Laika becomes the first mammal to orbit the Earth in outer space.

1957
Sputnik I is the first artificial **satellite** launched into Earth's orbit.

1961
Yuri Gagarin is the first human to go to space.

1965
Alexei Leonov is the first human to walk in space.

1969
Humans step on the moon for the first time.

1997
The Viking lander is the first spacecraft to land on Mars.

2012
Voyager 1 becomes the first spacecraft to travel into interstellar space.

AMAZING ASTRONAUTS

Many astronauts live and work in outer space where they conduct lots of exciting science experiments that we can't do from Earth. These clever, brave people discover new things all the time!

ASTRONAUT ADVENTURES

Astronauts are scientists and engineers trained to work inside and outside space vehicles. They have to undergo intense training before they're allowed to go into outer space. It can take them a few years to finish their training!

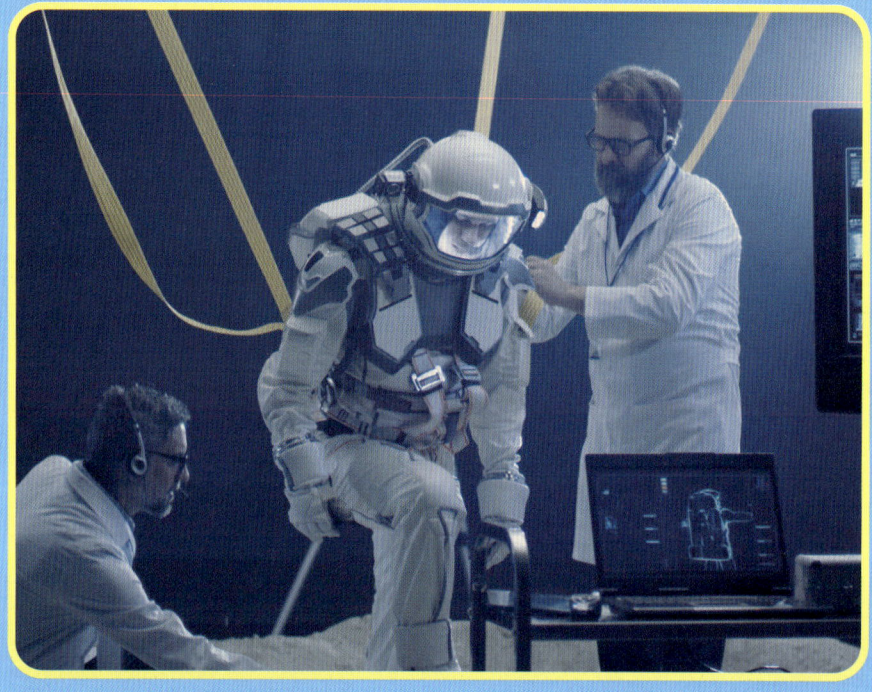

AMAZING SPACEWALKERS

Anytime an astronaut goes outside in space it's called a **spacewalk**. They usually last between 5 and 8 hours, depending on the job that needs doing. Astronauts are attached to the spacecraft to stop them from floating away, and they have a backpack with small jet thrusters so they can move around in space.

SUPER SPACESUITS

Walking in space isn't the same as walking on Earth! There are lots of risks with going outside the spacecraft. Scientists have had to make amazing suits to protect astronauts from the dangers of being in outer space.

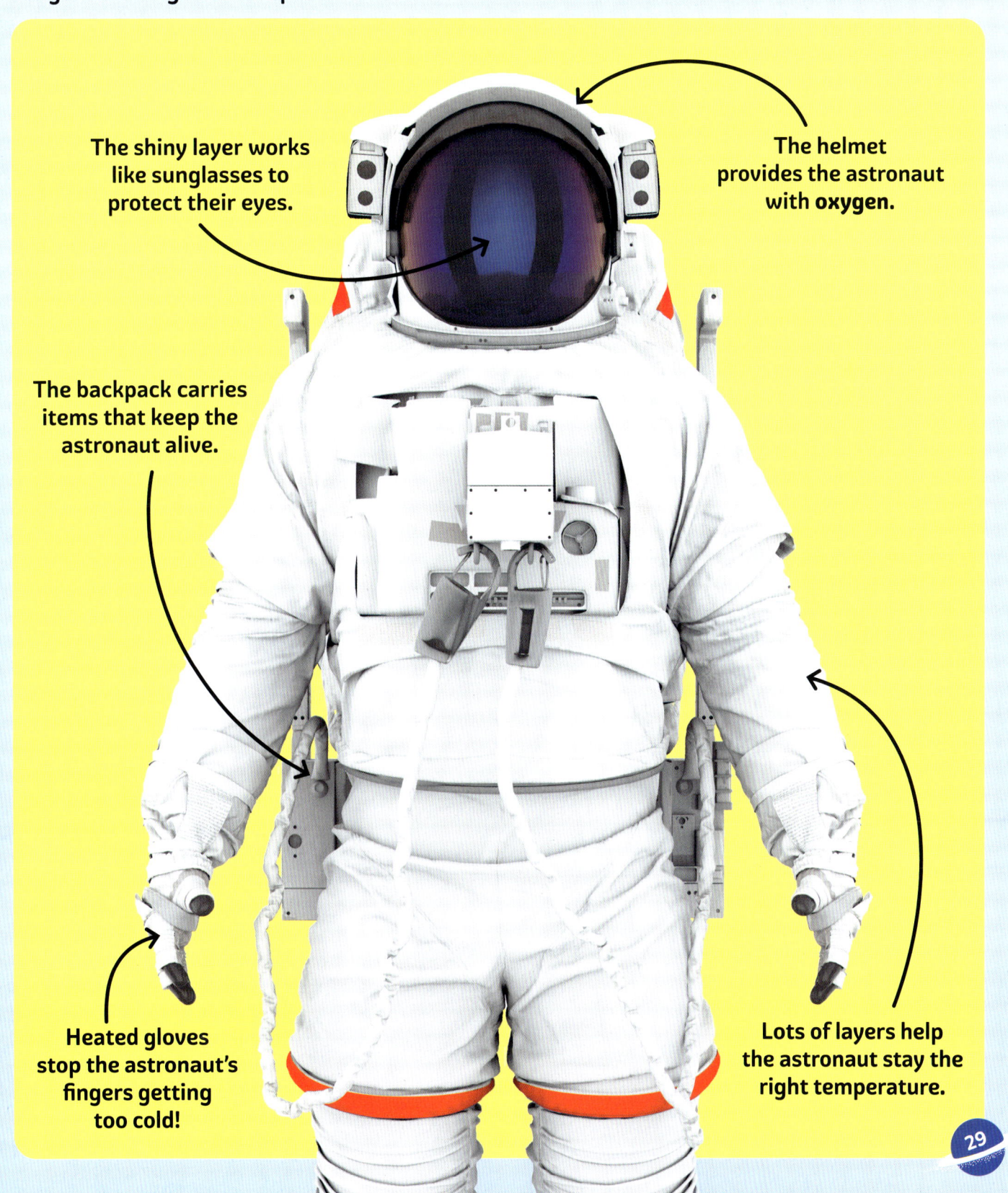

The shiny layer works like sunglasses to protect their eyes.

The helmet provides the astronaut with **oxygen**.

The backpack carries items that keep the astronaut alive.

Heated gloves stop the astronaut's fingers getting too cold!

Lots of layers help the astronaut stay the right temperature.

EXTRA-TERRESTRIAL EXPERIMENTS

Since space exploration began, scientists have been collecting and testing samples of material from objects in space, like the Moon. This helps them understand the type of rocks, chemicals and things that make each space object unique.

LIVING IN SPACE

As well as carrying out their own experiments, having astronauts living on the **International Space Station** is an experiment in itself! Because of this, scientists have made huge discoveries like how the human body changes after living in microgravity and how to grow food in space. All these things will help us prepare for future long-distance space travel.

HUMANS ON THE MOON

When humans journeyed to the Moon in the 1960s and 1970s, they collected samples of rock, soil, and dust to help them understand how the Moon and solar system were formed. The astronauts also left behind some experiments to collect information over time to send back to Earth, such as instruments to measure moonquakes.

COLLECTING COMET DUST

Scientists use space probes to examine **comets**. They've collected comet particles and dust during flybys which they use to understand what comets are made of. They've even landed a spacecraft on a comet before! It sent back information about the comet's temperature and photos of its surface which was brilliant data for scientists to examine.

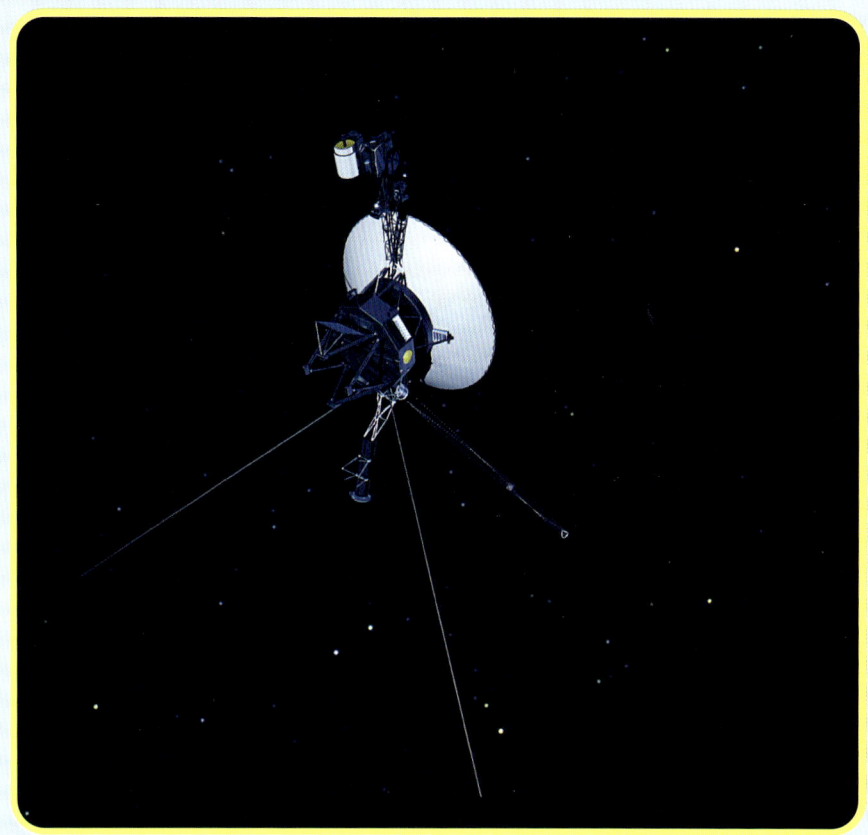

LIFE ON MARS

Humans can't travel to Mars yet, so scientists have sent **rovers** to explore the planet. They've tested the soil and pebbles, and found lots of minerals that we have on Earth, and signs that the planet used to have water. Rovers have also discovered that Mars has levels of radiation that could be dangerous for astronauts and will affect humans' ability to travel - in the future.

GLOSSARY

Aerodynamics – the study of how air moves around a solid object. The more aerodynamic an object is, the better it will fly.

Air resistance - an invisible force that tries to slow things down as they're moving through the air.

Astronomer – a scientist who studies outer space.

Comet - a large flying space object that's made of frozen gas, rock, and dust. As they get closer to the Sun, they heat up and appear to leave a glowing streak behind them.

Gravity - a pulling force that works across space. Objects don't have to touch each other for gravity to affect them. For example, the Sun, which is millions of miles away, pulls on Earth and the other planets and objects in the solar system to keep them in orbit.

International Space Station (ISS) – a large spacecraft in Earth's orbit which is used as a base for scientific research.

Interstellar space - the area of outer space beyond the boundary of our solar system., where our Sun no longer has any influence.

Orbit – the repeated path taken by one object circling around another object in space.

Oxygen - an invisible gas in the air that people and animals need to breathe.

Rovers - robotic machines designed to explore other planets.

Satellite – any object that orbits a planet. Satellites can be natural, like moons, or artificial (man-made items), like the communications satellites we use to send phone calls and TV signals, and for weather forecasting!

Space probe – a spacecraft which travels through space and sends data back to scientists on Earth.

Spacewalk – when an astronaut spends time outside the spacecraft while in outer space. They do this to repair spacecraft and satellites, and carry out scientific experiments.

Thrust - the force that pushes something in a particular direction. For example, the power of a rocket's engine pushes the spacecraft forward.

INDEX

A
astronauts 9, 16, 27, 28-29, 30-31

C
comet 31, 32

E
Earth 9, 10-11, 12-13, 14-15, 16-17, 18-19, 20-21, 22-23, 26-27, 28-29, 30,31
eclipse 14-15
engineers 26, 28

G
gravity 9, 17, 19, 30, 32

L
lander 27

M
Moon 14-15, 19, 27, 30

O
orrery 10

P
parachute 16-17
planets 8, 18-19, 21,23
probe 26, 31, 32

R
rocket 24, 26
rover 31, 32
ruler 7,14, 16,18-19, 21, 24

S
Space Race 26
spacesuit 29
spacewalk 28, 32
Sun 9, 10, 12-13,14-15,23

V
vacuum 9, 20

PICTURE CREDITS:

(Abbreviations: t=top, b=bottom, m=middle, l=left, r=right, bg=background)

Shutterstock: Albert89 27tr; Andrei Armiagov 30tr; Arkady Mazor 27mr; Artsiom P 31tr; Blue Planet Studio 28bl; Claudio Caridi 26br, 27br; FishCoolish Astronaut character throughout; Frame Stock Footage 28tr; Gorodenkoff 30bl; Joshimerbin 31bl; Klyaksun spaceship/rocket throughout; Muratart 15b; Nerthuz 29bg; Siberian Art 12b; Urvana 8b; Vadim Sadovski 23b. NASA Images: 2-3b; NASA/6900937 27mr; NASA/JPL-Caltech/University of Arizona 27bl. Wikipedia: Esther C. Goddard - Great Images in NASA 26tr; NASA / Peter Gorin / Emmanuel Dissais 26ml; Ria Novosti/Science Photo Library 27ml; Soviet (now Russian) Space Program via TASS (Dog Laika) 27tl.

Every effort has been made to trace the copyright holders, and we apologise in advance for any unintentional omissions. We would be pleased to insert the appropriate acknowledgements in any subsequent edition of this publication.